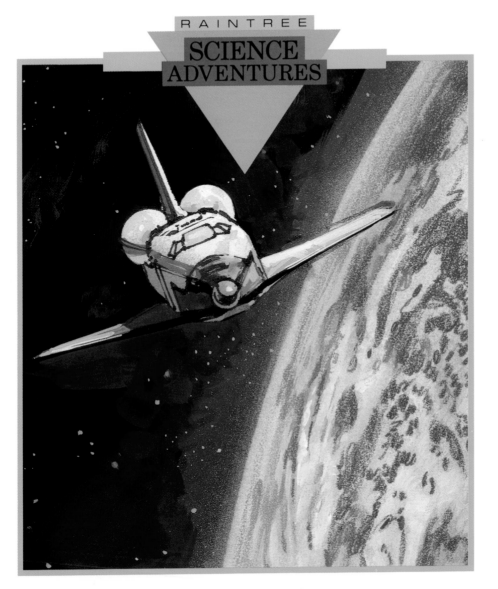

SPACE

Judith E. Greenberg and Helen H. Carey
Illustrated by Rick Karpinski

Raintree Publishers
Milwaukee

To Jason and Jessica — **J.G.**
To Virginia — **H.C.**

Editorial

Barbara J. Behm, Project Editor
Judith Smart, Editor-in-Chief

Art/Production

Suzanne Beck, Art Director
Kathleen A. Hartnett, Designer
Carole Kramer, Designer
Andrew Rupniewski, Production Manager
Eileen Rickey, Typesetter

Reviewed for accuracy by:

Gretchen M. Alexander, Executive Director
West 40 Educational Service Center Number 5
Northlake, Illinois

Scott R. Welty, Instructor
Maine East High School
Park Ridge, Illinois

Copyright © 1990 Raintree Publishers Limited Partnership

Library of Congress Number: 89-78292

1 2 3 4 5 6 7 8 9   94 93 92 91 90

Library of Congress Cataloging-in-Publication Data

Greenberg, Judith E.
  Space / by Judith E. Greenberg and Helen H. Carey;
illustrated by Rick Karpinski.
  (Raintree science adventures)
  Summary: Describes the nature of outer space, different kinds of rockets, and the planets of the solar system.   1. Manned space flight—Juvenile literature.   2. Rockets (Aeronautics)—Juvenile literature.   3. Planets—Juvenile literature.   [1. Outer space—Exploration.   2. Planets.]
I. Carey, Helen H.   II. Karpinski, Rick, ill.   III. Title.   IV. Series.
TL873.C37 1990 629.45   89-78292
ISBN 0-8172-3754-2 (lib. bdg.)

Before You Begin

This book takes *you* on an adventure in space! You will explore the world of an astronaut, learn what it's like without gravity, build a model spaceship, and be in the control center during an exciting space mission.

Your space adventure will let you take part in an experiment that will help you understand action and reaction. You will learn how it feels to live and work in space.

Knowing the words below will help you in your adventure.

astronaut   a person who is trained to be a space traveler

galaxy   one of billions of systems in the universe that include stars, gas, and dust

gravity   the natural force that causes objects to move toward the center of the earth or another body in space

hydrogen   a colorless, odorless gas

nuclear reaction   a series of changes in the structure of a nucleus, producing great amounts of energy

satellite   an object, either natural or human-made, that orbits a planet in space

simulator   a machine that imitates something that is real

space shuttle   a spacecraft that travels from earth to space and back

Now turn the page, and begin your science adventure!

Day 1 On a Space Mission

You are lying on your back strapped into your seat.
Above and all around you are dials, switches, and
control buttons. You have to remember what each one
does. You must also remember what to do if there is a
problem on the spaceship.

Suddenly, a red light flashes, and a buzzer goes off.
There is an emergency on board the ship.

The dial under the flashing red light shows you that
an air hose is broken. You must find the broken air hose
and fix it quickly. Otherwise, you must turn on the
reserve supply of air.

A voice from the control station on earth tells you
that if you use the reserve air, you will not have enough
to finish the mission. You will have to return to earth.
You must decide if you can find and repair the broken
hose in time.

You flip the switch to turn on the computer map of
the air-supply system in the spaceship. You can see the
break! You will be able to repair it quickly. The mission
is saved!

There is a loud pounding on the spaceship's door. You open it and see the smiling faces of your friends. You aren't in space at all! You are on the ground at a special training camp for future **astronauts.**

"The exercise you went through was part of a test to see if you have the right abilities for space travel," explains Astronaut Mission Specialist Karen Yatabe.

"Traveling in space is dangerous work. Astronauts must be brave and stay calm when something unusual happens. An astronaut needs a good mind and a healthy body," Ms. Yatabe tells the campers.

She adds, "Some people might think that you should have broken open the spaceship windows to let in air. Here's why that would have been the wrong thing to do."

She takes a paper bag and blows into it until it fills up with air. Then she pops the bag with her hand. The bag collapses because all the air inside it rushed out when the bag was popped.

"The bag took on a full shape when I blew air into it," says Miss Yatabe. "That shows that air has pressure. If you would have broken a window in the spaceship, the air inside the spaceship would have rushed out like it did from the bag. There would not have been any oxygen nor the right amount of air pressure in the spaceship that people need in order to live."

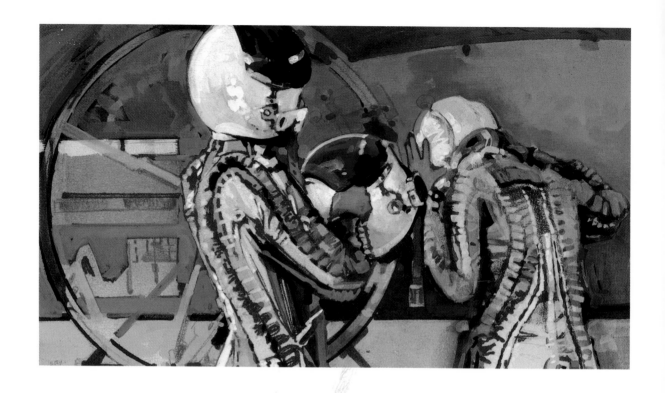

Day 2 Living in Space

The next day, you have breakfast with the other campers. Then everybody goes to the training center. You enter the training center through a series of doors. These doors keep the correct air pressure inside.

Inside the training center, several crews are training at the same time. Everyone is learning how astronauts live in space.

Your crew takes turns putting on a space suit. When it's your turn, you put on an undersuit first. The undersuit keeps astronauts cool because it has tubes in it that are filled with water. With some help from the crew, you get into your space suit. You put on a cap that astronauts call a "Snoopy cap." It has a microphone and headphones in it. A helmet is lowered over your head and fastened to a metal ring on your space suit. You pull on big gloves. Now you are ready to learn how to walk in space.

You have watched your friend Rudi training with a machine called the moonwalker. This machine shows what it is like to move around on the moon. The moon has less **gravity** than the earth. Even pressing down lightly on the floor with one toe caused Rudi to hop high into the air. You take a turn with the moonwalker. It feels very strange and different to move with less gravity than you're used to.

Next, you try out the spaceship **simulator.** In this machine you move in different directions. You go up, down, left, right, and forward across the floor. You spin and tumble just as you would in a spaceship.

Next, you watch your friend Sam in the manned maneuvering unit, called the MMU for short. Sam steers the MMU by pushing the hand controls on the ends of the armrests. There are places on the MMU that hold tools, cameras, and lights.

Astronauts use the MMU when they go outside a spaceship to make repairs on **satellites.** You practice with the MMU, too. On a space journey, everyone has to know how to operate all the machines. That way, if something happens to one person, the next person can take over.

At space camp, you spend some time in a room with no gravity. The members of the flight crew take turns cooking. When it's your turn, you heat and serve food in trays like airlines do. Straps hold the spoons and forks on the trays so that the utensils won't float away. Sam is still wearing the space suit and big gloves, so he must have only liquids. There is a drinking tube and a packet of liquid food inside the space suit.

In space, you might have to eat upside down. The muscles in your esophagus— the tube inside your body that goes from your mouth to your stomach—push the food down.

The soft drink that you will be having comes with a special cover. This cover keeps the fizz inside the can. Without this cover, bubbles would float all over the spaceship.

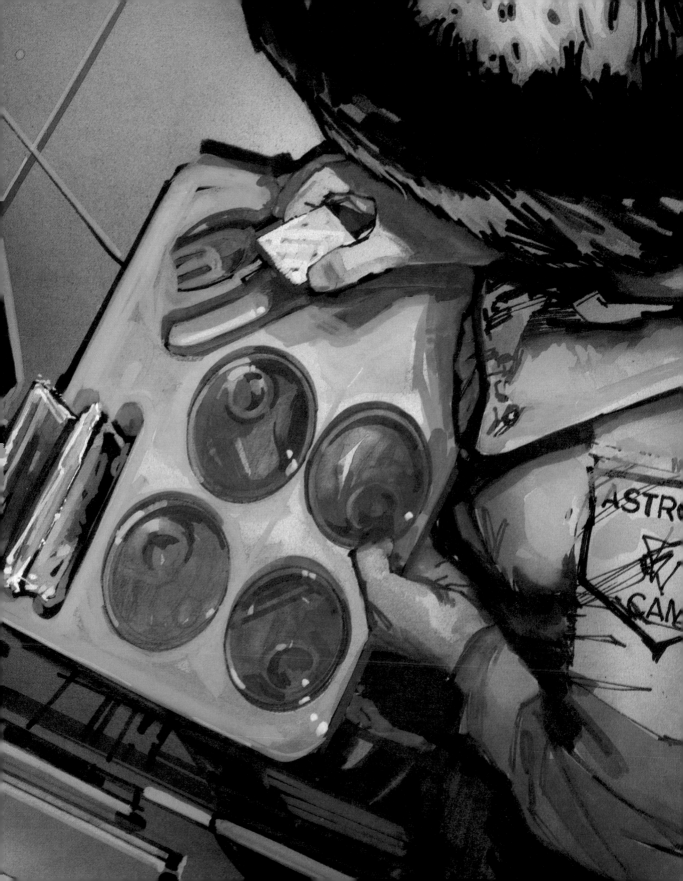

Later in the day, you ride an exercise bike. If you did not exercise during your space mission, the muscles in your body would get weak. This is because, in space, your muscles do not need to work to "fight" gravity. Astronauts need to exercise their muscles every day during a long space flight.

At the training center, you also learn what it is like to sleep in space. At bedtime, you get into a sleeping bag and strap yourself to a padded board. You can sleep standing up or lying down. Any way you adjust the board is comfortable because, in space, you weigh almost nothing at all. Heavy objects that you could not lift on earth also weigh very little.

Some crew members are still at work. You wear eye covers and earplugs so that you are not bothered by them. You don't need as much sleep as you do on earth. This is because, with weightlessness, you don't use as much energy.

When you wake up, you want to get clean. You use a wet cloth instead of taking a bath or shower. You cannot have a bath or shower because the water would float all over the spaceship. The spaceship's toilet stores body wastes until the ship can return to earth.

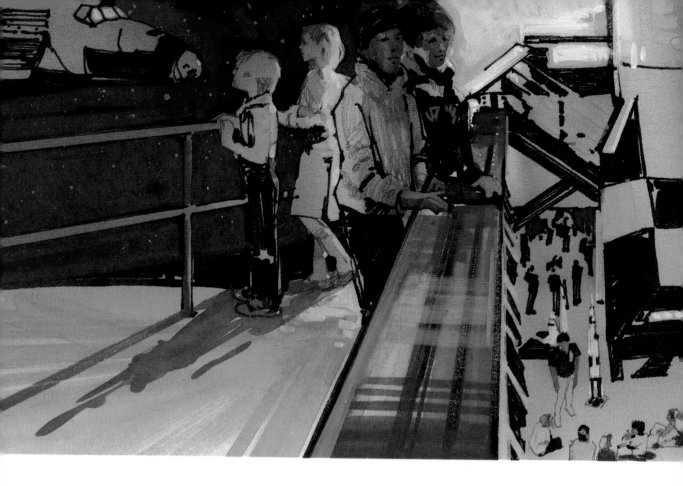

Day 3 Space Museum

"T minus 5 . . . 4 . . . 3 . . . 2 . . . 1 . . . 0. We have lift-off! All systems are go!" calls the voice over the loudspeaker. You hear the thunderous noise from the main engines. Then you hear the almost-deafening sound of the rocket boosters. The giant rocket with the **space shuttle** attached to it roars into space.

You wish you were on the shuttle, but actually you have been sitting on a bus watching a television. The TV was replaying the moment that every astronaut works for: lift-off into outer space. The bus is taking you to the space museum to see some of the world's first rockets.

At the museum, you see a model of *Sputnik I,* the world's first earth satellite. The Soviet Union put *Sputnik I* into space in 1957.

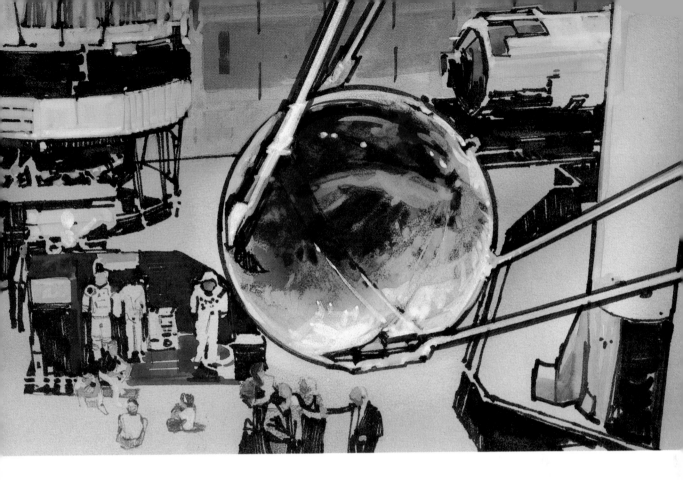

You see a model of the rocket that, in 1961, lifted the first American into space. His name is Alan Shepard. You also see a huge model of the rocket that launched John Glenn's spacecraft into orbit in 1962. Glenn was the first American to orbit the earth.

A guide shows you a model of a rocket engine and says, "During the Gemini IV mission in 1965, Ed White became the first man ever to walk in space. An engine like this helped put him up there."

You walk through a model of *Skylab*, the first American space station. *Skylab* proved that people can live and work in space for long periods of time.

The museum is filled with information about important events in the space program. You wish you could stay all day.

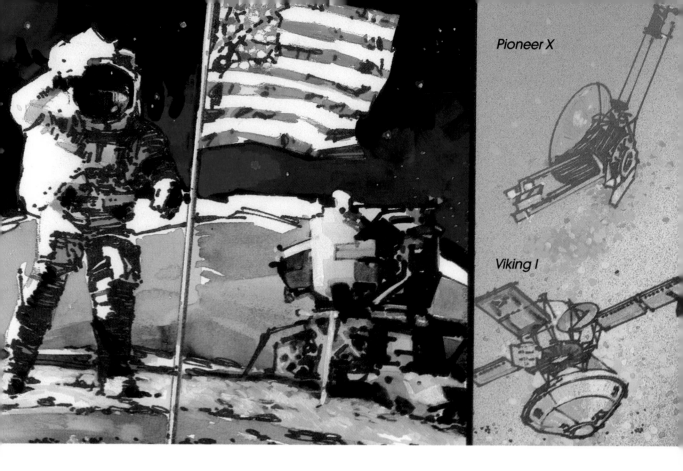

Pioneer X

Viking I

Moon Landing

Astronaut Neil A. Armstrong became the first person ever to walk on the moon. He landed on the moon in a small vehicle called a lunar landing module.

Space Probes

A probe is a spacecraft that is sent into space to gather information about other planets and stars. There are no astronauts on board a space probe.

The *Viking I* probe landed on Mars to photograph the planet's surface and study its soil and atmosphere. The *Voyager 2* probe photographed Jupiter, Saturn, and Uranus. *Mariner X* took photos of and transmitted information about Venus and Mercury. *Pioneer X* studied Jupiter and was the first spacecraft to travel beyond all the planets.

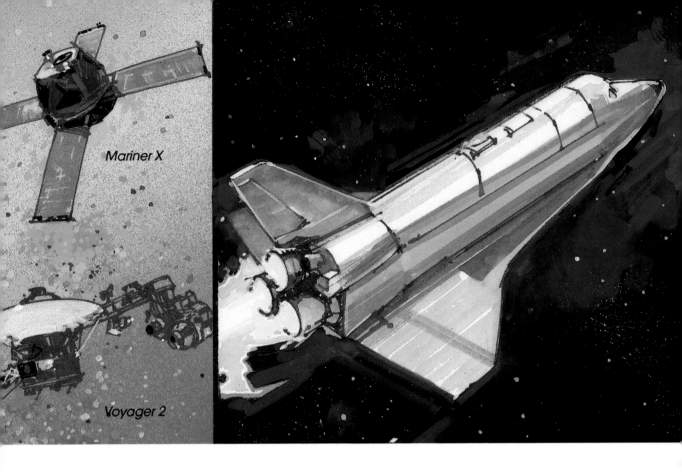

Mariner X

Voyager 2

Space Shuttles

Space-shuttle systems started being used in the
United States in the early 1980s.

The shuttle goes into space riding piggyback on a
rocket. The shuttle returns to earth by landing like an
airplane. The members of the shuttle's crew live in the
airplane section, which is called the orbiter. Shuttle
crews conduct many experiments while in space. They
launch new satellites and catch broken satellites. The
broken satellites are either fixed in orbit or brought
back to earth for repair.

On January 28, 1986, the world was shocked when
the shuttle *Challenger* shuttle exploded, seventy-four
seconds after its launch. All seven members of the crew
were killed.

Day 4 Spaceship Contest

The next morning, you meet other campers in the cafeteria at the space museum. Everybody is excited. The campers have just learned that there is going to be a contest to build model spaceships. Each crew will design and make a model for the contest.

Everybody in your crew has an idea for the model. You talk it over until a design is agreed upon. Later in the day, your crew will launch and fly your model spaceship in the contest. You hope your crew's spaceship will be the best!

Anthony and Mary draw a picture of the model your crew will make. Sam and Rudi pick out the materials needed to make the model.

Here is what your crew uses:
2 dinner-size plastic-coated
    paper plates
colored, see-through plastic sheets
felt-tip markers in different colors
scissors   tape

You follow these directions and build the spaceship:

1. Cut out a circle about four inches across in the center of one plate.
2. Lay the cutout circle on a piece of the colored plastic. Trace around it, making the colored plastic circle a little bit bigger than the circle from the paper plate. Cut out the plastic circle, and tape it over the hole in the paper plate.

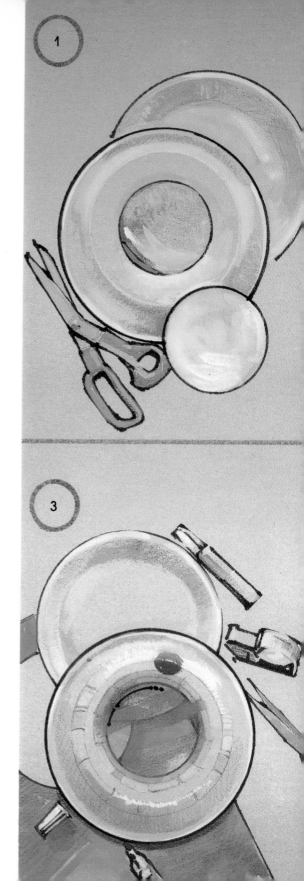

18

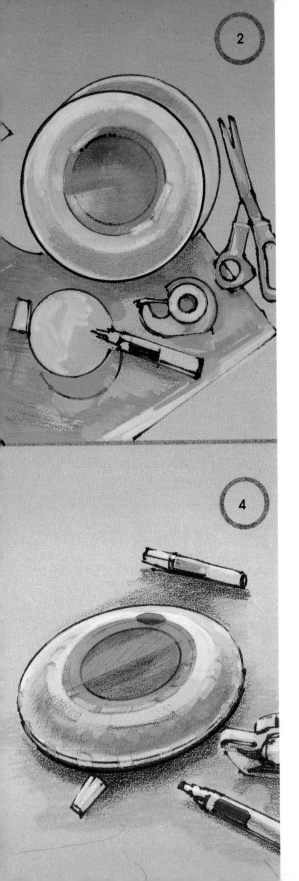

3. Draw windows on the top plate with the markers.
4. Use short pieces of tape to fasten the two plates together.
5. Decorate your spaceship with stickers, if you have them. Otherwise, draw some designs on the spaceship.

You test your spacecraft invention by tossing it like a Frisbee. It works well.

Then your crew meets with the others for the contest. Each crew launches its spaceship. Yours is the most colorful and flies the farthest!

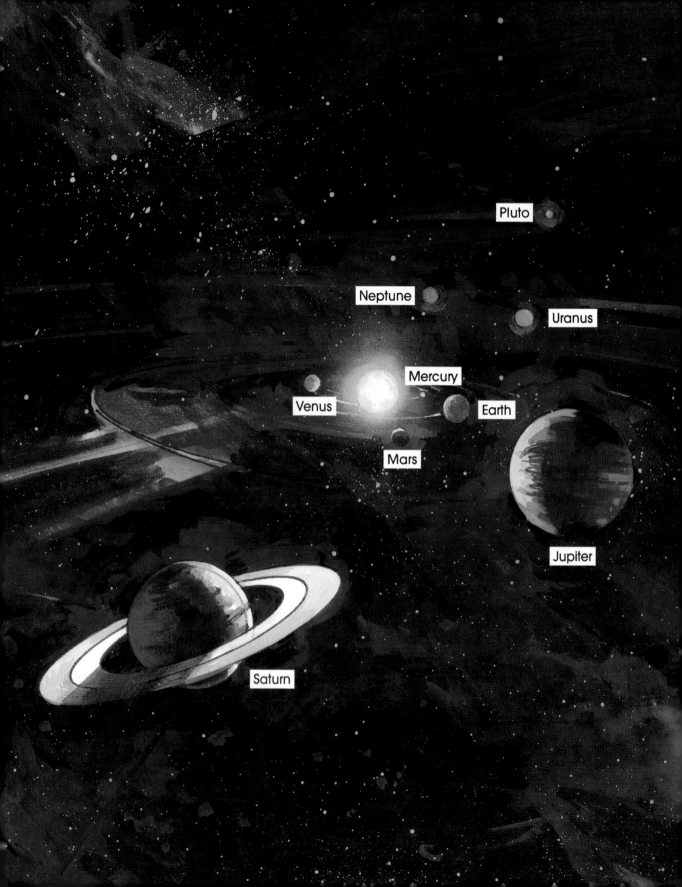

Day 5 Our Solar System

All the campers meet at the camp theater the next day. You are going to see a movie about outer space. The first thing you see on the screen is the blackness of space, dotted with millions of stars. All the stars that you see are part of the Milky Way **Galaxy.** In the Milky Way Galaxy, there is a huge star that is our sun.

From the movie, you learn that the sun is almost 865,000 miles (1,392,000 kilometers) across. It is 109 times larger than the earth. The sun is mostly made of **hydrogen**. The sun gives off heat and light.

Our sun and all the planets, including Earth, that travel around the sun make up our solar system. *Solar* means "of the sun." Each planet orbits, or travels in a path, around the sun at its own speed. Our solar system is just a small part of the Milky Way Galaxy.

The movie shows *Mercury*, the planet that is the closest to the sun. Mercury is the second smallest planet in our solar system. The surface of Mercury has bowl-shaped craters like on our moon. Mercury always faces the same side toward the sun.

Venus is the second planet from the sun. Venus has no water. The surface of Venus has volcanoes and craters.

Our planet *Earth* has both oceans and fresh water, air that can be breathed, and living things. Earth is the third planet from the sun.

Mars is only half as big as Earth. This planet has windstorms that fill the air with red dust. This planet has giant mountains and canyons. Mars is the fourth planet from the sun.

Jupiter, the fifth planet from the sun, is the biggest planet. Layers of dense clouds surround this planet. There is a continuous giant hurricane on Jupiter. If you look at the planet through a telescope, this hurricane can be seen as a red spot.

Saturn is the next planet from the sun. Saturn is ten times bigger than Earth. Rings made of billions of pieces of ice surround Saturn.

Uranus, *Neptune*, and *Pluto* are so far away from Earth that we know only a little bit about them. Uranus has thick clouds of gas. Neptune seems to be much like Uranus. Pluto is the smallest planet in our solar system. It is also one of the coldest planets.

When the movie is over, Ms. Yatabe gives everyone a long balloon and a small square of cardboard.

"Maybe one or more of you will become astronauts someday and travel to another planet in a rocket. We are going to use these balloons to do an experiment. The experiment will help explain how a rocket can go up into space," she says.

Following Ms. Yatabe's directions, you punch a hole in the middle of the cardboard with a pencil. Then you push the open end of the balloon through the hole. You blow up the balloon, holding the end tightly so that the air doesn't come out. You point the balloon toward the ceiling and let it go!

Your balloon rocket zigs and zags and zooms all over the room. Why does this happen?

(Write your guess on a piece of paper.)

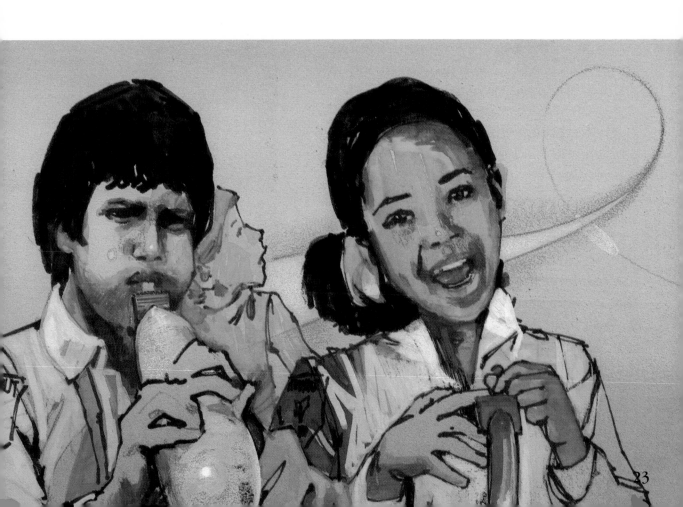

23

Ms. Yatabe explains: "About three hundred years ago, a scientist named Sir Isaac Newton said that for every action, there is an equal and opposite reaction.

"When you let go of the balloon, the air inside it came rushing out. As a result, the balloon moved in the opposite direction from the way the air was traveling. This is similar to the way a rocket works."

You look puzzled. Ms. Yatabe explains again, "When the air in the balloon pushes in one direction, the balloon pushes in the opposite direction. As the fuel in a rocket burns, the burning creates a downward force. The rocket then goes up into space."

"Even if Sir Isaac Newton didn't know about rockets, he knew how to make them work," you say.

That evening, you get to look through a telescope. You learn that most telescopes are tubes that have glass lenses at each end. Scientists called astronomers use telescopes to see, study, and photograph stars.

You learn that a radio telescope does not use lenses. It "listens" to radio waves that planets and stars give off. The radio waves travel right through clouds that can make it hard to see into space. Astronomers use radio telescopes to study planets and stars in any weather.

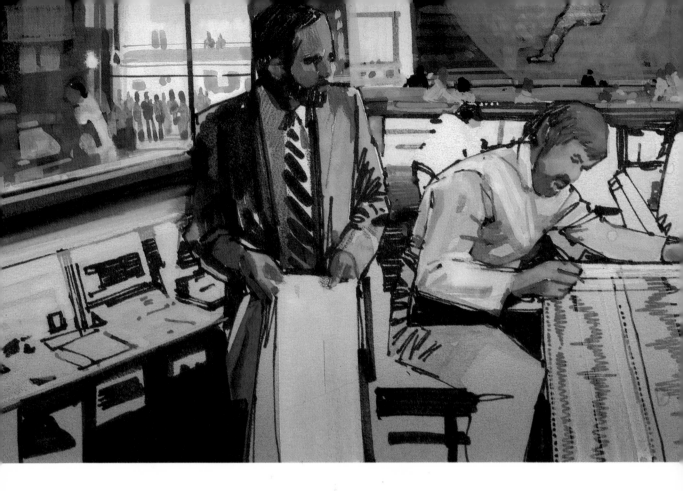

Day 6 Rescue in Space!

Today is a dream come true! You are going to explore space with real astronauts. Their space shuttle has been in orbit for five days. You will be on the ground in the control center. Using a monitor, you will be able to see what the astronauts see, and you will get to talk to them. They have been busy with experiments and taking photographs.

You walk into the control center. You sit in front of a monitor that looks like a television screen. You put on your headset. With the headset, you can hear the astronauts talking. The monitor in front of you allows you to see what they are doing. Another monitor shows you the view from the spaceship's window.

From the shuttle, Commander Drake talks to all the campers. He says, "I'm glad to have you with us on today's mission. You will explore space right along with us. I hope you see that being an astronaut is exciting and interesting. Does anyone have any questions?"

You ask a question, and it gets sent along to Commander Drake. You have asked if you'll be able to see a black hole today.

Commander Drake explains, "Black holes are invisible. One day we hope to have a space station that has special equipment to see black holes. Scientists have a theory that a black hole is a star that is old and dying. The star collapses and becomes smaller like a balloon that has run out of air. Its own gravity pulls everything near it into it, even light. We cannot see the dying star itself. The black hole is where the star used to be."

Rudi asks how stars get to be stars.

The commander says, "We think that bits of gas and dust are attracted to one another and begin to join together. They form a clump of material. Eventually, the ball of gas in the center of the clump heats up. The gas gets so hot that **nuclear reactions** occur. The heat is so great that the outer layers of the clump give off light. At this point, a star like our sun is born."

Commander Drake would like to answer more questions, but a flashing red light has come on in the spaceship. You see the light on your screen, too.

"That red light means we've reached an important point in our orbit. Our spaceship is near the communication satellite that we have come to repair. Keep your eyes on the monitor, and you will see me during my EVA," says Commander Drake.

In astronaut camp, you learned that *EVA* stands for extra-vehicular activity. This means that Commander Drake will go outside the spaceship to work.

You see the commander go through an opening in the spaceship. His tools are strapped to his suit with Velcro. A long line called a tether attached to his space suit keeps him from floating away into space. Small jets attached to the back of his suit move him toward the satellite.

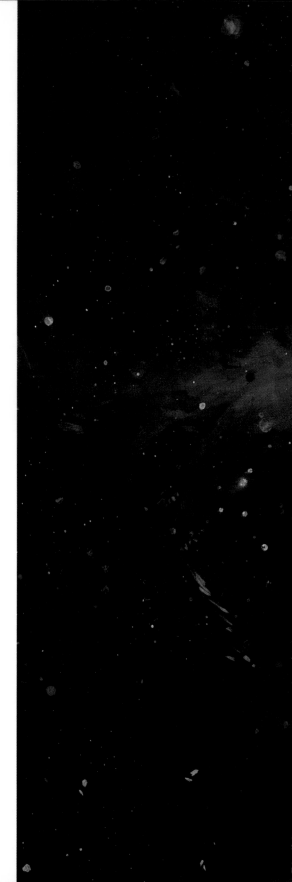

You watch your monitor. You wish you were in space, too.

Suddenly there is a crackling noise in your headset. People are shouting. It's hard to understand what the astronauts are saying.

Susan yells in your ear and points to the screen, "Look, Commander Drake's tether broke."

"He'll be lost in space. Why doesn't he use his jets to get to the spaceship?" asks Rudi.

"What is he waiting for?" you ask.

All the campers are scared!

Everyone listens as Commander Drake's calm voice comes over the speaker. "A small meteorite, which is a piece of solid matter, hit the tether. The tether snapped like a twig. I'm going to use the jets and get back to the ship."

There is a big sigh of relief from all the campers. You watch, but nothing happens.

Commander Drake slowly drifts along in orbit with the spaceship and the satellite. "The jets aren't working!" says the commander.

Just then you remember Sir Isaac Newton's idea—action and reaction. "Maybe the commander should push against the satellite," you say. "Then he'll go in the opposite direction. He'll go toward the ship."

As if he has heard you, Commander Drake pushes against the satellite as everyone watches. The gentle push sends him close enough to the spaceship to grab the tether. He's safe! The rest of the mission is carried out, and the astronauts begin the journey home.

Astronaut camp is over. You will remember it always. Now you know for sure that you want to be a space scientist when you grow up. You might be an astronaut and live in a space station. You might be an astronomer and explore faraway stars with a telescope. Maybe you will have a job inventing probes for space exploration.

What kind of science adventure would you like to go on next?